Rescuer Beware

An Effective Guide for a Healthy Rescue

By
Dr. Shirley Moore

© Copyright 2010 by
The Mobile Prayer Warriors Ministries

No part of this publication may be reproduced, stored in a retrieval system, or transmitted in any form or by any means. All rights reserved except for brief quotations to printed reviews without the prior written permission of the publisher.

RESCUER

BEWARE

Train up a child in the way he should go, And when he is old he will not depart from it.

Proverbs 22:6 NKJV

Introduction

When the people of God rescue the ungodly who are holding on to their evil or wicked ways; what is actually being rescued is the evil or wickedness and the rescuer is perpetuating the cycle of disaster. Until man finds discomfort in his situation he will not seek change. Some pain is necessary for maturity and development. However, God has not appointed us to inflict pain but He allows the pains of life to promote a mindset for correction.

There are many parents today that are trapped in the cycle of rescuing their children far into their adulthood. The children of the Christian, like those of any others, have to make their own choice as to who they will serve. The laws of the scriptures apply to them as well as to anyone else. In other words God will not change His standards for our children. The choices that they make will affect their lives for good or for evil. Often the rescuer's life is damaged or altered significantly by their commitment to those who are holding fast to their ungodly ways. In this hour of disaster upon the earth many are rescuing others. There is nothing wrong with helping those who cannot help themselves. There is a thin line between helping and enabling. Rescuing is sometimes done out of a false sense of responsibility. It is often done out of a sense of loyalty when rescuing loved ones. We rescue at times due to our compassion for others thinking it is the humane thing to do. In our rescuing we want to be very sure we are not creating dependency or enabling negative behavior.

Some rescue out of pride while others rescue out of guilt. There are as many reasons for rescuing others and most of them good. However if you rescue a fool in his folly you will need to rescue him again and again until he learns to avoid folly. How will they learn if there are no consequences to their actions? We don't want to call anyone a fool but if they fit the definition (a fool hates instructions) it is safe to say they are serving folly and the cycle of disaster in their lives will not end until they relinquish their foolish ways.

Persons that require constant rescuing but refuse to accept advice should not be rescued. The soul of the rescuer will be tied to the sins of those who they hold onto, while they, are holding onto sin. The mind, will and emotions of those who are trying to lead and guide their children out of the ways of destruction are often intertwine to the point of the rescuer being held captive. Often the rescued is leading the rescuer into death, lost and destruction by proxy.

We will not give up on the salvation of our children and neither will they drag us fearfully into the pits of the hell they choose to explore. One must maintain their own freedom in order to set others free. The first step to setting the captive free is to uncover why they are bounded. It is the captive that must make a decision for freedom. The captive may need those very pains of life we are so anxiously trying to rescue them from, to be truly set free. The pain of discomfort will help them to determine they want and need to change. Knowledge has to be embraced and wisdom reverenced by those who need to be rescued. The knowledge of the truth and consequences attached to good and evil begins with the fear of the Lord. Fear enough to know that disobedience will cause great suffering. Embracing freedom from sin and its traps on earth will start the journey to eternal freedom with the Lord. We must teach our children these truths, abandon them or spend the rest of our lives in the rescue cycle. They must be rescued from their own evil ways and not just from the consequences resulting from bad decisions and sin.

For the ways of man are before the eyes of the Lord and he ponders all of his goings. His own iniquities shall take the wicked himself, and he shall be holden with the cords of his sins.
Proverb 5:22-23

Poverty and shame shall be to him that refuseth instructions: but he that regardeth reproof shall be honored.
Proverbs 13:18

The wisdom of the prudent is to understand his way: but the folly of fools is deceit. Proverbs 14-8 *Whoso rewardeth evil for good, evil shall not depart from his house.*
Proverbs 17:13

A man of great wrath shall suffer punishment: for if thou delivers him, yet thou must do it again.
Proverbs 19:19

Definitions

1. Fool: A person without sense of judgment; persons who act unwisely; stupid or silly person.

2. Folly: A being foolish, lack of sense; unwise conduct. A foolish act or practice, or idea; something foolish, a costly but foolish understanding.

3. Rescue: Save from danger, evil or harm. Rescue means to save a person by quick and forceful action from immediate or threatened danger or harm.

4. Wicked: Twisted thinking, sinful, bad, cruel, worthless, malignant. Doing nothing with what you have.

Un-Bewitching Saints Children

In this year marked 2011 in the month of February the world is experiencing trouble like never before. There are wars and rumors of wars. Disaster is coming from winds, rain, earthquakes, floods, and fire. Then we have HIV, STDs and viruses in common with these. We also have cancers, heart attacks, lupus and other prevalent diseases spreading rampantly in the lands. These things have their peaks and valleys and therefore do not command the constant attention of those not directly affected by them. However, there are un-natural, personal disasters that have compounded the injuries of the world. Divorce, greed, immorality, violent tendencies, the search for pleasure and escape by seeking earthly values has caused inordinate affection in the minds of the children of this world.

For the purpose of this writing the term "children" refers to anyone of any age that is having to or depending on their parents or others to survive because they are making bad decisions. It also refers to those whose life styles are negatively impacting their parents in any way. This is not an age group however those most effected are between the ages of 13-35. This is a wide range but the iniquity, in this day and age, is wide spread and is influencing the Body of Christ indirectly through the unsaved children of the Saints. The enemy has worked a work in the children to cause the saints of God to take their eyes off of their God and to chase their children for fear that the disastrous life styles they are living, leading them to death, loss and destruction.

We have put forth too much effort to safeguard our children from consequences and thereby enabled them to remain dependant on others and to oppose themselves. We find ourselves doing everything we can to save our children. We came to the Lord because we realized we were undone and on our way to destruction. We gave thanks to God for what He had already done for us and worshipped Him for who He is. We served Him because we loved Him and appreciated Him for our relationship. He drew us and we came, eventually.

Now that we find we have children that are not coming into the knowledge and appreciation of God as we desired that they would. We are interacting (not interceding) with God on their behalf. We find ourselves bargaining, scheming and plotting to reach God for them. Let us examine ourselves, become humble, repent and turn from our wicked ways so that our God will hear from heaven and heal our land. (see Deuteronomy 28). We have bribed and pleaded with our children often to no avail. Now it is time to allow God to do the work of a Father in their lives to the saving of their souls.   In this fast moving world filled with higher and higher demands day by day we as parents have learned to pacify our children to get them to be content while we try to provide for them. We are often caught up trying to save them from earthly discomfort rather than from hell fire. Weather we are providing their meals, doing their laundry or working for a living, we have often had to throw something together for them quickly so we could get on about the business of survival. We had little time to see deeply into their lives. Life was training the parents of the 60s and 70s to move quickly or be left behind. We therefore trained our children to expect quick responses or risk being neglected and unfulfilled. They learned to feel entitled to everything without earning much, if anything. In the process the parents were trained to respond to the children without really listening or understanding. We were accustomed to shoveling several things at them in rapid sequence until they stop their age appropriate methods of petitioning for our intervention. As infants they are crying, then toddlers whining, teens yelling, and then demanding until finally as adults depending and expecting to be rescued.

We sit back and wonder what went wrong. Why did they not grow up and become independent and productive adults? The Bible says train up a child in the way he should go and when he gets old he will not depart from it.

Let us Pray:

Lord, have mercy on our souls for our souls were not properly fixed on You when we were raising these children. Now they are old and they are not departing from our training, but the training is not producing the intent and purpose of our hearts. Therefore, we can only cry for mercy now, that You might help us to break the cycles of destruction and turn us and our children around that we might once again do things

Your way. Let the time be redeemed for our families. Our Father, You are great in forgiveness and Your mercy endures forever. Be merciful unto us. Amen.

There are times of rest and peace written of in the Scriptures. We are not in a time of peace as the world calls peace but we often find ourselves discontent and disappointed because we can not obtain the peace that the world seeks and has always expected on earth. We will be at complete rest and peace when Jesus returns and the author of confusion is bound for good. Meanwhile we endure hardship as good soldiers as the Word encourages us to do. We do not endure hardships for the sake of endurance but for the prize of victory at the end.

We must rest in Jesus Christ our Lord and Savior and the knowledge of His victory. Jesus set the victory ahead of us and then set us to go to the victory and to obtain it in His name. He paid the price of our victory in full just as one would pay for a ticket for another to travel to a destination. The traveler must go and say, " my travel has been paid for by Jesus Christ and I am here to start my prepaid journey at His expense."

We must battle against the enemy of our souls (mind, will and emotions) to stay on the pathway of righteousness to the finishing of our journey in life on earth. We have the peace that Jesus spoke of in Luke 14, which surpasses understanding. It surpasses understanding because we experience His peace in the time of war, in the midst of the storms of life. Christians are promising and being promised heaven on earth with little or no conflict touching them.

There is not a life recorded in the Bible that gives us an example of this trouble free existence. As we look at Peter, Paul and John, John the Baptist, Steven and Jesus our personal example, we see sorrow and grief throughout their lives. We can look at these and see that man's days are short and full of trouble. Most of man's driving force of frustration comes out of the false belief that he is entitled to heaven on earth. Man purposes to live as if the devil is already stripped of his ability to influence the world for evil. Man is often frustrated because he is not appearing to win the battle against his enemies of poverty, sickness, violence, confusion, distraction and captivity. Man has busied himself fighting all of these elements both natural and spiritual in quest for wealth that will give rest on earth.

Our children see our struggles and hear our prayers. They have learned to expect a fairy tale world and are disappointed when they experience discomfort in this world. A false expectation of a troubled free life from God causes a loss of faith. We have not been taught for the most part and have not taught, that we are here to do the will of God. Children should be taught from birth that they are to seek God's will and to pursue the pathway of righteousness all the days of their lives.

Neither of the Apostles, disciples nor Jesus himself, who is the Lord of the Armies, ever obtained the peace that man seeks on earth. They did not see all of the victories they fought for while in the flesh on earth. Some battles are won today but the results are not seen until years later. Jesus fought and won a way for us out of no way. Those that were inspired to live the scriptures and record them for us endured hardship and suffered loss to follow Christ. They served the purpose and cause of God while on earth.

Hebrews 12: *"Wherefore seeing that we also are compassed about with so great a cloud of witnesses, let us lay aside every weight of sin, which doth so easily beset us, and let us run with patience the race that is set before us. Looking unto Jesus the author and finisher of our faith. Who for the joy that was set before him endured the cross, despising the shame, and is set down at the right hand of the throne of God. For consider him that endured such contradiction of sinners against himself least ye be weary and faint in your minds."*

While dealing with our children we have come into some situations that have a tendency to bring shame. Beloved we must be able to endure the shame of the cross just as Jesus did. We cannot throw in the towel (faint) or give up because of the contradiction of sinners we are facing in the generations present and coming on. Our way of thinking must be changed from peace as the world offers peace to peace as Jesus offers peace. No matter how much money we acquire, money cannot buy peace. It can buy happiness until a circumstance comes up that money cannot fix and happiness will be fleeting. Happiness depends on what is happening at the time and what happens next. Joy comes with the peace of Jesus that surpasses all understanding. Man has inadvertently made peace itself an idol in that many are seeking a world free from trouble. There is still and adversary that has a little time left to influence the world for evil. He employs death, loss and destruction as his handymen. The Saints of God have been deceived to a point and let's not go any further in the deception. We are not on this earth to establish our own utopia by any means and neglect the teaching of the Gospel of Jesus Christ to our families by example. The example some of us have been demonstrating says, we have to meet the demands of the world or we will die or be poor. We have become a servant to the will, demands and ways of the ungodly. Our actions demonstrate that we are full of fear and anxiety over obtaining a standard of living acceptable or respected by our peers evident by the things we can acquire.

We are placing our trust in gimmicks, rebellion (witchcraft) half-truths and false values. These are the principles our children are now comfortable with. They do not need our God to reproduce what we have produced. We have in many ways, without intent disregarded the laws and will of God in our interactions in the church, jobs schools and homes.

How can we promote God when we are doing everything contrary to His will and His ways? How can our children be excited and trusting of God when we show by our actions we do not trust Him?

The reader might be saying by now that yes others do that but not me. Let us look at ourselves very close and see if it is not true of most of us, well meaning Saints of the true and living God. Man has lied on the Holy Ghost, to the Holy Ghost and spread the lies amongst themselves until they now are well accepted as truths.

One lie is that when you accept Jesus as Lord and Savior all of your troubles are over. The truth is that you have obtained a very presence help in the time of trouble. Another lie is that Jesus came and died so man can go to heaven without living according to the Word of God under mercy and grace. The truth is God said, "be ye Holy as I am Holy" and at another place He says, "be ye perfect as I am perfect." Still another place He says, "I give you both the power and the will to obey.

To live Holy is to live in line with the Word and will of God for each individual. The individual purposes will then fit jointly together into the Body of Christ and be obedient to Jesus, the anointed one as the head of all things.

The Body of Christ is seeking something that was never recorded of Jesus and that is a life of pleasure and comfort. Some die feeling cheated of success as though they failed because they did not achieve the world's definition of success. The Body of Christ often finds itself striving like the body of Satan to obtain worldly goods. Ministers are presenting the Gospel as a commodity to be peddled to the highest bidder. The children are seeing this. They grow up to do the same thing or reject the Gospel entirely because of the false teachings and corruption. The pleasure of the people of God must be to become living sacrifices to do the Will of God on this earth to perpetuate holiness.

Do not let the devil trick you right here. Please, please do not allow the enemy to carry your mind away to think that this writing is teaching against prosperity for the Saints. It is not, there is nothing wrong with Christians being wealthy and prospering even as their souls prosper. Let the writing be truthfully interpreted to say that the love of money is the root of all evil. Ask yourself, how much do I love money? Where are my priorities? What will I do to obtain money? How much will I sell my birthright as a Christian for?

The devil is subtle, backsliding is slow and man's heart is wicked. If one thinks he stands let him examine himself honestly in the pure light of the eyes of His Father to make sure he stands properly. As one moves on in life and has their needs met the next move is to have their desires met and then we move to the love or lust for the persons, places or things deemed to be most important in life. Can we truly say that as Saints of God we always demonstrated righteousness to our children and others that are looking at our lives.

Let us pray:

Lord, God help us to see ourselves as you see us and as others see us. Help us not to misrepresent You or ourselves. You know the purpose and intent of our hearts. Yet if our purposes are pure and our intent be of righteousness, but we yet error in our ways the outcome will still be contrary to both Your will and ours. Thank You Father for correcting us. We will to please You and to do what is right in your sight. Teach us the truth of our ways and show us how to be in Your perfect will at all times that we might truly be salt and light to those that observe us in Jesus name we ask. Amen.

In today's world one needs money to survive, one desires money to be empowered and independent, to achieve comfort and mobility with options. One loves money to be powerful, or to be admired while exercising power, authority and control over persons, places and things. That is what Satan wanted.

Satan is working to promote his formula for success on man (love x money) = power. The evil one seeks to impart into every person the love of money to secure a place for his self in the form of pride and lust. The intent of Satan's heart is recorded against himself in **Isaiah 14:13:**

For thou has said in thou heart, I will ascend into heaven, I will exalt my throne above the stars of God. I will sit also upon the mount of the north; I will ascend above the heights of the clouds. I will be like the most High.

Satan is not sitting above God but he does appear to being sitting above some of God's stars. Satan uses "I" five times in two verses. One must watch oneself to see if "I" is becoming the most used word in ones hearts. I have to earn a living, I have to pay bills, I have to provide, I have to make sure, I want, I need, I have to have, I will not take this or that. Watch out for "I" statements. As we use "I, me, my and mine, we are often stepping into pride.

We say things to our children like: this is my house, my car, my food, my money and when you get your own then.... What is the message we are giving them? Is it not one of pride and selfishness placing things as more important than the children?

Therefore children will place things as more important then people. When one teaches responsibility one must be sure to instill teamwork and the sense of belonging to a family, class, church, community, or country. Those that rebel the most and are often the most defiant are those that are made to feel rejected or outcast from the general assembly and those that feel entitle because they never had to earn anything. The "I" syndrome is of the results of selfishness which leads to pride which come before a fall. (Proverbs 16:18.) When man starts to speak and say in his heart "I" in a selfish or prideful way, he is being influenced by Satan. He will soon become a member of his body with satanic values while sometimes perpetrating Christianity.

The body of Satan is composed of persons under the influence of the I, me and my syndrome. Satan's body is composed of those that run to evil, wicked and unrighteous inclinations to obtain gain at the expense of others. Those of the Body of Christ have to be watchful for the subtleties of the spirits of Satan. Understand this Satan has lined up an army that releases evil into the world. The children are deceived into seeking a never-ending party.

The Body of Christ is to release what they have freely received. Christians are to receive the love of God as fruit of the Spirit of God. God is love John 3:16. The fruit of the Spirit of God is love characterized by joy, peace, kindness, longsuffering, meekness, self control, gentleness and faith Ephesians 5, Galatians **5:24**: *And they that are Christ's have crucified the flesh with the affections and lusts. If we live in the Spirit, let us also walk in the Spirit. Let us be not desirous of vain glory, provoking one another, envying one another.*

Are we walking in the love of God that corrects and speaks the truth to all it encounters? When our children see us interact with others and experience us on a daily basis what is the picture of love they see? How are we defining love in action toward them and our fellow man? Is love a false sense of responsibility, guilt or obligation? Is the love picture we paint one that takes the time to correct, listen and assist with problems? Are we guilty of being in such a hurry in this fast lifestyle until we say far too often, "It is easier if I just do it myself." Day after day the child grows up and allows it to be easier and easier for you to do it yourself while they grow irresponsible and lazy.

There will come a day when there is no one to do for them and they will be inadequate at living. The Body of Christ is commanded to walk in love and to influence others to do so. The body of Satan is commanded to fulfill the lust of the flesh and to influence others to do so. The scripture ask how can two walk together unless they agree? In another place it is written; if they are not for Me they are against Me… Still another place that says if they do not gather, they scatter. Yet another Scripture says nothing comes to a double minded man. He is unstable in all of his ways. God also says, "I rather you be hot or cold, if you are lukewarm, I will spit you out of my mouth." Then it is written, choose ye this day who ye will serve, If Baal be god then serve Baal, if God be God then serve God.

We run into conflict in the households when our folks, sometimes children, start to act out in a manner not becoming to the teachings of God. Sometimes it is because the spirit in which we do things as well meaning as we might be does not resemble the Spirit of God. The fruit of the Spirit patience often flies out the window and is replaced with impatience and lack of tolerance for the growth process of others. Our children often meet frowns and disapproval and our fears of their failures. Our fears will not promote confidence in those we influence. Those around us respond to the spirit we release into the atmosphere rather than what we say.

Tone, facial expressions and body posture speak louder than any words we might use. I asked the Lord, how shall I go back and redo all of this now that they are adults. His answer is one event at a time. Slow down and be slow to speak and quick to hear. Pay attention. Produce the fruit of My Spirit only. Take the time to be Holy. I reflected back on what it means to be Holy. It is to be on one accord with God for His purpose and His will to be done. When God's will is done through a person all of the time, their environment will change for the good of all.

In life we must make many decisions. Before we go to the mountaintop there is a valley of decisions we must enter into. In the valley we must make the decision to be doers of every fresh Word God gives us. Do as He says quickly.

We made a decision for Christ as to determine where we would spend eternity. Once that decision is made one must continue to enforce that decision daily. Our decision was to allow Him to be our daily Bread. The will of God has to be done every day in order for us to get the desired results in our lives. We must tell our children we will follow the Lord and seek His will in every aspect of our lives including what we do or do not do for them.

The Christian must keep in mind that how we act, react, fulfill commitments, interact with others, handle vows or contacts will reflect Jesus in us. We are in covenant with God to be Christ-like on the earth.

Some say that children are a mirror image of their parents. Well, they are a combination of their parents and their ancestors. There are many personalities and traits that can pop up in a child. People are carelessly mating these days.

The Scripture says do not be unequally yoked with unbelievers. It also asks what business does the light have with darkness? The word of God says we are in the world but not of the world. God also instructed us to come out from among them and be separated. Some are justifying going any and everywhere doing about any and everything. If you are not on assignment from God to go into the darkness to expel the darkness you might want to avoid the darkness and prefer your Christian brothers and sisters. When among sinners be a light that shines in the darkness. Those of the Body of Christ must maintain a separate stance from the body of Satan. This is a stance denoted by what one does or does not lend the members of their bodies to. What ever one purposes in their heart is evident by the fruit they bare.

One of my biological sisters and I were talking on the phone concerning our children. We always include the Lord in our conversations because we are one with Him. Often times the Lord in His grace will give revelation to us or insight into His word for answers and help us in our times of need. On this particular morning the Lord began to show us how we were not following Him as closely as we thought we were and needed to in order to receive the desired results we were seeking. God showed us how we had actually been led by the spirit of another other than Him. God had already spoken to me days earlier that I needed to withstand the spirit in operation in my children.

I pondered that word for a while and even acted on it and felt devastated. The Lord gently said you are being directed and controlled by the spirit that is not of me that is influencing your children to influence you. I remembered God saying to me once before, you are following your children and where are they going? They were certainly not going anyplace I wanted to go. However, I began to realize I was being led by the decisions they were making good or bad. I was still responding to their cries as if they were still bottle babies. We have trained them and they have trained us to respond without consulting God for directions or help. The Lord showed us how we were not allowing them or assisting them to seek Him. They did not truly see themselves going to God except through us. We had become their god instead of Jesus Christ. Children of God, there are some things that have been going on for centuries that have hindered us from following Christ. Let us look at ourselves closely to discern the truth. We must hold fast to that which is good and rid ourselves of all that is not good. Let's look at the body of Satan, which is composed of those who are unrepentant and walk in unrighteousness, ungodliness, transgressions, and are trespassers. We the Body of Christ cannot be partakers with them by taking on the characteristics of their fruit. The Lord said you will know them by the fruit they bare. This includes our children. We must love them but reject any of their ways or practices that are ungodly.

In the records of the kings sometimes the next king followed their father and sometimes they did not. Even though they were the sons of their father some did evil like their fathers some that had evil fathers and did good. It is the same today, some of our children are following after God and some are not. As my sister Deborah and I discussed our children on that morning our concerns were the flaws in their characters otherwise called sin. We wondered why they were not practicing what we had literally preached to them. We were both licensed and or ordained to preach and teach the Gospel of Jesus Christ. We began to reflect on the reality that many of our sisters in the Lord were experiencing the same problems with their children. We lived holy lives before them, took them to church, prayed for them and over them, proclaimed the word of God and fought for their souls. Yet, these now, young adults for the most part, showed no signs of turning their lives totally over to God. They would defend God and tell others about God when the conversation came up but they still struggled to make a true and lasting decision for Christ. The fruit they produced was not coming to maturity. On this particular morning the Lord began to show us that our children were following us but not Him.

He gave us insight to understand that most of them did not know Him for themselves only through us. At first sight this may not look too bad. However, the Lord showed us that they were actually influencing our lives and we were reinforcing their dependency. They would do foolish things and we would rescue them. When things went so wrong for them until we could not rescue them we would be desperate. We would seek the Lord the more on their behalf. The Lord wants us to seek Him for others. However, they need to learn as soon as possible that they most depend on God for themselves. We can be in agreement with others for deliverance but everyone has to be in agreement. We learn to seek for others to know the Lord in the pardon of their sins.

Our children are still looking for the quick and painless relief which does not cost them a thing. If you rescue a fool in his folly, you will have to rescue him again. These young adults knew we knew God and they had seen God work on the behalf of the family but they did not reverence God, nor did they fear Him. They wanted to do some things that they knew would not please God and were comfortable doing so for the most part. It felt somewhat like they were confident that their mothers knew God and therefore they had immunity. They behaved like the "spoiled spiritually rich". Mind you, we are not monetarily wealthy by any stretch of the imagination. However, God has provided for us and they have witness Him make ways out of no way. David said he was young and then old but never saw the righteous forsaken or his seed begging bread.

Our offspring enjoyed the fruit of our labor in more ways than one. They were receiving from God through us and not through a relationship with Christ. We were also moving in the office of the Holy Spirit trying to lead them, guide them and teach them all truths and bring all things to their remembrance. God refused to be handed down indirectly. Everyone has to make a direct connection to Him for themselves. People anointed to lay hands on others can do so and release the power of God into that person for healing or deliverance. Yet, all those that seek the Lord must be willing to come into a new way of life through the appropriation of the shed Blood of Jesus Christ for the remission and washing clean of all sin. Our children were being compelled by devils to commit sin just like all other sinners. Satan could no longer influence us directly to do evil so he came at us through our children.

It is not that we were sinning but we were distracted by our children's crisis. Our lives were being dictated by their bad decisions. The devil would set sin baits and trap them legally, financially or health wise and we would go into the trap with them and fight their way out. Meanwhile our ministry would go lacking. We were always looking back trying to get them to follow. One cannot wholly follow the Lord while looking back at anyone. Our health would some times suffer and our bodies showed signs of duress. Our finances were always under siege by them or because of them. They wreak our cars, we bailed them out of jails, bought them food, paid their bills, set them up in businesses that failed, bought them cars. We drug them to church and so on.

The Lord was being merciful toward us because we diligently sought Him on the behalf of our children. We began to realize that no matter how bad things looked our God was showing mercy and grace because we asked Him to. We also realized that He would put them in the furnace of affliction if He needed to and it would still be because of His grace, mercy and love.

The dross and tin must be burnt off. We must release our children to God as well as all the others we pray for. We cannot instruct God, neither can we faint because of the crying of those in their afflictions.

When bad things happen or they look bad by sight, know that the Lord is working it out for our good. Keep trusting and praying and move out of God's way. He will bring their salvation into being in due time. Meanwhile do the work God has called you to do. Get into the center of His will and pray for His will to be done. It is His will that all come to repentance and be saved, especially our children. We are in line with the Word of God to pray for the lost. However, we must remember to direct them to God in love and withstand the influence of Satan through them.

They must be led by the Spirit of God to know what to do, how to do it and when to do it. There is no set way to respond except in obedience to God. He only knows the answer for every problem. There are no two people alike and no two solutions the same. The adult children were practicing all manner of sin and some of their children were following suit. The wages of sin is death. We wanted to show love, we were fearful of death and the lost of their souls. We finally got some consolation when we remembered the Scripture that says God is faithful to complete the work He has started in us. God had started a work in each of them, since they had all received the Lord in their youth. They are the seed of the righteous and therefore our children are delivered from every destruction. We continue to stand on their behalf but now we stand in the gap instead of being the gap. We have learned not to sponsor sin by running to their rescue without consulting God on what to do if anything. The old cycles had to be broken in love. They would get into binds or trouble of some sort and we, the parent, trained from their birth, would jump and run to their rescue leaving God and wisdom behind.

We really believed we were suffering for God and acting out of love for our children. When our loved ones are in danger, fear often grips our souls and we respond in fear and disbelief in the promises of God. Actually we were acting in fear of the power of Satan over them. Therefore we were being governed by the spirit of fear and Satan was activating fear against us through our children. Fear is not faith and all that is not faith is sin. We must remember that David and his army had been out fighting the enemy and while he was gone all of their children and wives were taken into captivity. David cried until he could not cry any more. He then came under attack by his men. He had to encourage his own self, but before he made a move he consulted God for instructions. He obeyed God and recovered all. We might have been praying but we would have likely been on our way after the enemy before we received instruction from God. We need help from heaven to recover all. When we seek to help without first seeking God we often cause the crisis to linger on and on. When we refuse to follow the instructions God gives us, we risk great loss or long delays in the manifestation of victory.

We think we are waiting on God but He just might be waiting on us to come to terms with what He has already commanded. Victory is already set for the people of God. We must use obedience and be led by the Spirit to activate the victory given to us by the shed Blood of Jesus Christ our Lord and Savior. God has ordained all members of the Body of Christ to be a working part of the Body to cause it to fit jointly together, to grow and maintain spiritual health. All sin sick souls are a disease to the body of Christ and cannot be allowed a place.

There are benefits from being born into families who are of Christ. However, everyone must know Jesus for themselves and therefore become connected directly to Him. We can not be born into His body from our mother's womb. Everyone has to be born again by making their own decision to make Jesus LORD of their lives.

One must be willing to give up the world's ways and take up the ways of our Father which is in heaven. To be born again is to become a new creature in covenant with the Lord to do the things that glorify God and His purpose for mankind. To be born again is to join the army of living God to hate evil and to love righteousness. The Lord spoke to us and said pray for their souls as if you were praying for their natural lives.

Let us pray:

Father God in the name of Jesus and by His shed Blood we ask you to save our children and the Saints children all over this land. Help us as parents to focus more on their soul salvation than on their lives. Father we pray that they all come to repentance and be saved without spots, blemishes or wrinkles in their hearts. Father help us to stand in the gap for others and not in the way. Help us to walk by faith and not by sight thereby bringing honor to You. We repent of being fearful and acting without consulting You first. Give us the wisdom and knowledge to win these souls and many more. Help us to spread the good news to all of those that are perplexed concerning the manifestation of the salvation of their loved ones. Let us be filled with Your Spirit and be governed by the mind of Jesus Christ. Father give us this mountain called family. Thank you Father Amen.

The Lord gave us a word saying, they are attached to you, let them remain attached, leave them but make preparation for them to follow. Establish new rules and conditions for them to obey. Follow me and lead them to deliverance. The word was given and we can instruct them as Paul did when he said follow me as I follow Christ. The Lord is saying to us follow Me and they will follow you. This is surely what we thought we were doing but instead we were following them as they were following our enemy. We could see the enemy working in them but we could not see him controlling us through them.

They spent their money on alcohol, entertainment, cigarettes, parties and friends. We brought their food, paid their rent and utilities. The more we rescued them the more they needed to be rescued; They fulfilled the lust of their flesh. We financed them while they were doing it. We tried to make life as pain free for them as possible. They ran out of gas and minutes on their phone which they needed to survive to let them tell it. We paid for the gas and the minutes and they spent money with no thought of tomorrow or necessities. They lived by the ways of the world governed by their own will and the lust of their flesh.

Galatians 5:17-21 says that those that do these things shall not inherit the kingdom of God. The kingdom of God is peace, righteousness and joy in the Holy Ghost. In their present state, our children could not inherit the kingdom of God. They did not practice the ways of God. They were always looking for a quick fix, a short cut or a way around the rules. The wages of sin is death. Satan promotes pain for punishment and torture to those who fail to do his will or fail at their assignments to destroy others. God on the other hand allows pain to warn us when we are in danger of death, loss and destruction. As parents we try to prevent our children from experiencing pain which is the natural consequences of wrong doing. Pain is necessary in life to cause those that are stubborn to turn around and go in the right direction.

As human beings we do not and cannot know without God, what to do to counter act the attacks on families today. There are hordes of demons coming against us but we know that if God be for us He is more than the world against us. As Blood bought believers in Almighty God we know to keep our eyes on Him and He will direct us to victory. We must make sure we are standing correctly in the gap for all mankind that is walking in the darkness. They must see Jesus through us. They must not see a fairytale God and expect heaven on earth. We must learn to say no to them when God says no. When they are in the furnace of affliction we must let them know that they need to call on the name of Jesus to be saved. Everyone has to come into the knowledge of who God is and that He will set the rules for every covenant with every person. He does not barter or trade. He says you do this and I will do this. Follow Him and He will lead each one who follows to safety.

Saints of God we must not follow after our loved ones as they follow after Satan. If we are not sober minded and watchful, Satan can reach through our children and pull us out of God's will by influencing them for evil. We must love God and trust Him always knowing that He loves our people more than we do. It is God that saves and gives mercy to those that call on Him.

As we give our offering unto the Lord and pay our tithes, the Lord blesses us. He expects us to be good stewards over what He blesses us with. However, if the enemy is controlling those that are influencing the spending of the children of God's money than whose kingdom are we really promoting? We ought to be sure to pray over whatever we give out. It needs to bring in a harvest to our God. The Lord blesses us to be a blessing. It is not against God's will for us to feed the hungry and cloth the naked. It is His will for those that are of Him to demonstrate love by helping others to help themselves. It is unfruitful to promote dependency, riotous living and bad habits. However, wisdom tell you that if a person is an alcoholic to the point that if he does not get alcohol he will die, he needs to be kept alive until he accepts Jesus. He that wins souls must be wise. For each circumstance presented by each individual one must hear the wisdom of God and act accordingly.

Let us be careful that we do not promote evil to the left or to the right in our desire to do good. Many seek the help of the church in times of need but do not seek Jesus Christ the husbandman of the church. One has to be connected to the fountain of life in order for their true hunger and thirst to be quenched forever. Man has many needs, even the rich has needs that only God can meet. When we allow Jesus to truly be the head of His own body in our eyes, He will lead us and guide us to all victories.

Let us as the Body of Christ refuse to participate and promote the suffering of those being used by the devil as he tries to solicit the help of the Christians to sustain bad habits and spread death, lost and destruction.

Establish ground rules by the leading of the Lord. Watch and pray, reprove all things, hold fast to that which is of God. Produce the fruit of the Spirit only so that those around you might be partakers of that which is good and brings life. Be patient with all men and with yourself as the Lord completes the work He has started in each of us.

Let us pray:
Lord contend with those that contend with us and save our children. Keep them from the stranger's ways. Keep our hearts from fearing evil and help us not to do any harm in our efforts to do good. Let every plot and plan of Satan be defeated in our lives by the Blood of Jesus that was already shed for the remission of sin and made atonement for all mankind. Let all come to repentance and be saved, especially our children and our children's children down to the generation until the return of Jesus. We pray in agreement with the Saints of God everywhere that we will be effective warriors against the hordes of Satan and his own defeat will be manifested in every life that a Blood bought Christian comes into contact with. May the time be redeemed and the years that the locust, canker worms and grasshoppers have eaten be restored. We caught the thief with Your help Lord and now he has to restore sevenfold what he stole. We forgive those that have sin

against us and receive forgiveness for all of our sins. We do not complain about the duration of test that have revealed our hearts, nor time spent in the process of deliverance. Instead we thank You Father for Your patience with us and Your love and kindness toward us as You mold us. Lord we thank you for completing the work you have started in each of us. We thank you Father for the precious Blood of Y'shua (Jesus) and Your response to our use of His name in asking for Your continuing peace that surpasses all understanding. Father help us to meet the next crisis in faith in You. Let us not stagger at the promises but with confidence that you have heard our prayers. As You beckon us to come up higher we come up to Your throne of grace to receive mercy in our times of need. Take us Father from glory to glory in Y'shua's (Jesus) name we pray. Amen.

For information on other ministries provided by The Mobile Prayer Warriors Ministry such as Prayer Shut Ins-Deliverance Ministry- Community Outreach, Training Seminars, books, CDs and ministering the Word of God. The Mobile Prayer Warriors consist of Intercessory Prayer Warriors Evangelist, Ministers, Teachers and Youth Ministry please feel free to contact us at:

mpwim@yahoo.com
prayingagain@yahoo.com
dj_mpw@yahoo.com.

Manufactured by Amazon.ca
Bolton, ON